How Animals Evolved

HOW FISH EVOLVED

by Elizabeth Andrews

DiscoverRoo
An Imprint of Pop!
popbooksonline.com

WELCOME TO DiscoverRoo!

This book is filled with videos, puzzles, games, and more! Scan the QR codes* while you read, or visit the website below to make this book pop.

popbooksonline.com/ev-fish

abdobooks.com
Published by Pop!, a division of ABDO, PO Box 398166, Minneapolis, Minnesota 55439.

Printed in the United States of America, North Mankato, Minnesota.

102023
012024

THIS BOOK CONTAINS RECYCLED MATERIALS

Cover Photo: Shutterstock
Interior Photos: Shutterstock, Getty Images, Science Source, Alamy
Editor: Grace Hansen
Series Designer: Laura Graphenteen

Library of Congress Control Number: 2023939059

Publisher's Cataloging-in-Publication Data
Names: Andrews, Elizabeth, author.
Title: How fish evolved / by Elizabeth Andrews
Description: Minneapolis, Minnesota : Pop!, 2024 | Series: How animals evolved | Includes online resources and index
Identifiers: ISBN 9781098245436 (lib. bdg.) | ISBN 9781098245993 (ebook)
Subjects: LCSH: Evolution--Juvenile literature. | Fishes--Juvenile literature. | Fishes--Evolution--Juvenile literature. | Biology--Juvenile literature.
Classification: DDC 591.38--dc23

*Scanning QR codes requires a web-enabled smart device with a QR code reader app and a camera.

TABLE OF CONTENTS

CHAPTER 1

WHAT IS EVOLUTION?

Evolution is the scientific **theory** that all life on earth has a common ancestor that lived billions of years ago. Living things **developed** from earlier versions of that ancestor. The changes that happened took millions of years.

WATCH A VIDEO HERE!

Betta fish are also often called Siamese fighting fish. They are popular pets.

DID YOU KNOW?

Almost all scientists believe in evolution because it is testable.

Changes that help an individual survive might be faster legs so it can outrun predators.

Evolution happens because of natural selection. This process occurs when one individual of a **species** has different **traits** from others. Sometimes the differences make surviving easier for the individual. As they survive, they pass on their traits to their young. Over time, more and more individuals of the species are born with the new and helpful trait until it is normal.

Evolution is how fish eventually moved onto land. It is how birds began to fly and mammals started giving birth to live young. It is how humans began walking on two feet.

Ancient fish swam in different bodies of water than fish today.

CHAPTER 2

WHAT ARE FISH?

Fish are animals that live in fresh water and salt water. They come in all shapes and sizes. There are at least 34,000 **species** of fish. Fish may have scales or slimy skin. Most fish breathe oxygen through their gills underwater, but some

LEARN MORE HERE!

Parrotfish (bottom center) *have a strong bite to help them eat coral.*

fish can breathe air. The three main types of fish are bony, **cartilaginous**, and jawless fish.

Salmon are born in fresh water and move to salt water.

Since there are so many types of fish, they don't all give birth the same. For example, a blue shark gives birth to live young. Clown fish, salmon, and bullhead sharks lay eggs. Their young are born outside of the mother.

DID YOU KNOW? The smallest fish species is less than half an inch (1.3cm) long. The largest species can be more than 40 feet (12.2m) long.

Not all fish that exist have been discovered. The ocean is very deep. Scientists haven't explored all parts of it yet. Unknown and amazing fish could be lurking below waiting to be discovered.

CHAPTER 3

ANCIENT FISH

Tracing the history of fish is more difficult than other animals. Their fossils are hard to come by. Many ancient fish and some modern fish do not have bones. Their bodies are made from different materials that don't last.

EXPLORE LINKS HERE!

Ancient fish shared waters with the animals that would evolve into insects.

The oldest fish is called *Metaspriggina* from 518 million years ago (MYA). This animal was the ancestor of all **vertebrates**. It had gills, two eyes, and a flexible rod down the back of its body. Oxygen was absorbed from water through the skin. At this point in fish evolution, gills were used only to take in food.

DID YOU KNOW?

Many fossils that survived are likely to be buried deep under the ocean floor. They are hard to get to.

Fish who lived after bones evolved were more likely to become fossilized.

As fish became larger, they needed more oxygen and food to stay strong. Gills were covered in blood vessels. Oxygen started being absorbed through them. Fish **developed** open mouths to eat instead of getting food through gills.

Jawless fish have round, open mouths. Scientists consider *Haikouichthys* to be the first jawless fish. These were worm-like animals with the beginnings of fins. Jawless fish dominated the ocean from 500 MYA to about 300 MYA. These fish fed off living and dead animals. Ostracoderms were jawless fish that developed armor on their heads and bodies. The armor protected them from **predators**.

Haikouichthys *had backbones and a distinct head.*

Jawed fish evolved from jawless fish. Jawed fish had mouths with upper and lower parts that moved on their own. Jaws let fish catch bigger prey. *Entelognathus primordialis* lived 420 MYA. It had cheeks and jaws. Fish developed more fins, paired nostrils, and strong bodies made from cartilage or bones.

FANTASTIC FINS

Fins are used by fish for many things. They help with forward motion and steering. They are also used for protection on some fish. Fins may be for display. They attract partners to have young with.

Scientists study fins because they were the first example of paired limbs in animals. Fins would eventually evolve into limbs, such as legs and arms for animals who live on land.

Not all fish have skeletons made from just bone or cartilage. Some fish have both.

CHAPTER 4

MODERN FISH

Jawless fish evolved into fish with armor. Then came **cartilaginous** fish followed by bony fish. Cartilaginous and bony fish are still swimming in Earth's waters today. As time has gone on, fish have **diversified** into both weird and beautiful animals.

COMPLETE AN ACTIVITY HERE!

The longhorn cowfish lives in coral reefs.

Rays live on the sea floor. Their mouths are on the bottom of their bodies to catch prey.

Cartilaginous fish include sharks, rays, and sturgeon. They do not have bones. The cartilage their bodies are made from is strong and flexible. Sharks often have rows of sharp teeth. They have highly **developed** hunting skills. They can sense the electricity of their prey. Sensing electricity helps with **navigating** waters too.

The oldest shark fossil was from 480 MYA. Sharks have survived five mass extinctions that killed many other animal species.

Today, 90 percent of fish are bony fish. They began overtaking the waters on Earth about 340 MYA. The largest group of bony fish are called ray-finned fish. Their fins are webs of skin filled with bony spines. The spines are a bit like fingers.

Surgeonfish only have one dorsal fin.

FISH FINS

caudal fin: most important fin for movement, propels fish forward, only fin attached to the spine

dorsal fins: **stabilizes** the fish to keep it from rolling, helps with turns

anal fin: stabilizes the fish to keep it from rolling, helps with turns

pectoral fin: helps with side-to-side and up-and-down movement

pelvic fin: stabilizes the fish, helps with up-and-down movement

Male seahorses carry the eggs from female seahorses. They keep them until it is time for them to hatch.

Bony fish have developed into interesting **species**. Unlike some animals with matching limbs, fish fins may grow from different parts of the body. Some fish may have more fins than others. They also may have different sorts of scales. A pufferfish looks completely different from a sea horse, but they share common ancestors.

Pufferfish swallow water and air to make their bodies expand.

Through their 500 million years of existence, fish developed exciting skills. The electric eel has a special organ composed of electric cells. It can make

Eels are ray-finned fish. They usually have one fin that runs along their body.

Anglerfish can grow to be four feet (1.2m) long.

a large electric charge for protection or hunting. The lanternfish has body parts that make light to help it see in dark places. Fish are still evolving new skills and into new species today!

MAKING CONNECTIONS

TEXT-TO-SELF

What part of the evolution of fish surprised you the most? Please explain your answer.

TEXT-TO-TEXT

Have you read any other books about animal evolution? What did the evolution of those animals have in common with fish evolution?

TEXT-TO-WORLD

There are thousands of species of fish in the world. With the help of an adult, look up information about your favorite one. Write a short paragraph about your favorite fish.

GLOSSARY

cartilaginous — made of cartilage. Cartilage is soft, bendable tissue.

develop — to grow over time.

diversify — an increase in the variety of animals that are different from one another.

navigate — to find a way through.

predator — an animal that lives by hunting and eating other animals.

species — a group of living things that are very much alike.

stabilize — to help keep steady.

theory — an explanation for how things work and why things happen.

trait — a quality or feature of something.

vertebrates — a group of animals that have a bony or cartilaginous spine.

INDEX

DiscoverRoo!
ONLINE RESOURCES

This book is filled with videos, puzzles, games, and more! Scan the QR codes* while you read, or visit the website below to make this book pop.

popbooksonline.com/ev-fish

*Scanning QR codes requires a web-enabled smart device with a QR code reader app and a camera.